Water quality and GIS

Dr. Hemant Pathak

ISBN: 1482380331
ISBN-13: 978-1482380330

DEDICATION

All true praises are only for Shri Sainath Maharaj, the guardian of the worlds, who blesses the humanity with his creative ability.

Dr.Hemant Pathak

Contents

Foreword

Water Quality and GIS; provides a unique insight into the problems our planet faces in terms of water quality and quantity, and what to do about it. This is the only books expressed comprehensive and interdisciplinary focus to hydrological understanding with the multidimensional approach.

This book made of 06 years consistently research on water resources, makes it ideal source for students, teachers, industrialist, water experts and environmentalists.

This book provides an essential guide to researchers, it offers: various aspects of water; on the challenges and experiences in present scenario.

Simply explained, Water Quality and GIS is an important book for all who wish to make a difference in how to plan and manage our water resources.

Until and unless Water, that magical substance from which all life springs forth, is essential to the very existence of every life form on earth. The role of water in the living organism has not changed since life's first creation in salt water billions of years ago.

Dr. Hemant Pathak

M.Sc. (Gold medalist), Ph. D.

Assistant Professor of Engineering Chemistry

Indira Gandhi Govt. Engineering College,

Sagar, MP, India

Glossory

GIS

Acronym for *geographic information system*. An integrated collection of computer software and data used to view and manage information about geographic places, analyze spatial relationships, and model spatial processes. A GIS provides a framework for gathering and organizing spatial data and related information so that it can be displayed and analyzed.

GIS coordinate

In Survey Analyst for field measurements, the single coordinate for a survey point that is the best overall representation for that survey point's location, defined by one or more projects. Feature geometry is always linked to the GIS coordinate.

GIS Data Reviewer

An application used to manage data quality control, visually check data and run batch checks for attribute and geometry defects. Defects are recorded in an error table that can be used to resolve errors and verify corrections.

GIS server

The components of GIS Server that host and run services. A GIS server consists of a server object manager and one or more server object containers.

GIScience

Abbreviation for *geographic information science*. The field of research that studies the theory and concepts that underpin GIS. It seeks to establish a theoretical basis for the technology and use of GIS, study how concepts from cognitive science and information science might apply to GIS, and investigate how GIS interacts with society.

Global analysis

The computation of an output raster where the output value at each cell location may be a function of all the cells in the input raster.

Global Check method

In Survey Analyst for field measurements, one of two ways to apply the Coordinate Out of Tolerance command. The Global Check method searches for coordinates out of tolerance within the whole survey dataset.

Global functions

The computation of an output raster where the output value at each cell location may be a function of all the cells in the input raster.

Global mode

A navigation mode during which the camera target is always at the center of the globe.

Global Navigation Satellite System

The Russian counterpart of the United States Global Positioning System.

Global polynomial interpolation

In Geo-statistical Analyst, a deterministic interpolation method. The interpolated surface is not required to conform to the sample data points, and the method does not have standard errors associated with it.

Global Positioning System

A system of radio-emitting and -receiving satellites used for determining positions on the earth. The orbiting satellites transmit signals that allow a GPS receiver anywhere on earth to calculate its own location through trilateration. Developed and operated by the U.S. Department of Defense, the system is used in navigation, mapping, surveying, and other applications in which precise positioning is necessary.

Global spatial data infrastructure

A global framework of technologies, policies, standards, and human resources necessary to acquire, process, store, distribute, and improve the use of geospatial data across multiple countries and organizations.

Graphical user interface

A software display of program options that allows a user to choose commands by pointing to icons, dialog boxes, and lists of menu items on the screen, typically using a mouse. This contrasts with a command line interface in which control is accomplished via the exchange of strings of text.

Water Quality and GIS

1.1 Introduction

Water is one of the most essential natural resources for eco-sustainability and is likely to become critical scarce in the coming decades due to increasing demand, rapid growth of urban populations, development of agriculture and industrial activities. Water is valuable natural resources that essential to human survive and the ecosystems health. Water are comprises of coastal water bodies and fresh water bodies (lakes, river and groundwater).

Water is the most important natural resource which supports all life forms on earth and has no known alternative. Earth has abundant water due to the presence of the water cycle on it but even then most of it is unfit for human use and consumption. Out of approximately 71% of water present on the surface of the earth, less than 3% is potable and easily accessible to human beings.

Water is a unique substance, because it can naturally renew and cleanse itself, by allowing pollutants to settle out (through the process of sedimentation) or break down, or by diluting the pollutants to a point where they are not in harmful concentrations.

Besides this Increase, population and its necessities have also lead to the deterioration of surface and sub surface waters. Human never cared for maintain the quality of water; and always destroyed the rivers, lakes, and oceans.

Contamination of surface water has become a major challenge to environmentalist in the rapid developing countries. By mapping water quality using the decision support system like GIS, it can be useful for taking quick decision based on graphical representation.

Water also acts as a purifier in our bodies. If enough water isn't consumed, one is unable to properly flush out their kidneys and/or liver, and the colon is unable to expel bowels properly and completely thus keeping unhealthy toxins in the body. As a result, the toxins are able to make its way through the human body causing poisoning and spreading infections. The characteristic of water can be categorized into three namely physical, biological and chemical. These characteristics are used in water monitoring program.

Water sustains life for humans, animals and plants. People need water for basic everyday activities like drinking and cooking, but water is also very important for the fuelling of agriculture and industry, and plays an important role in the nature of national economies.

However the supply of freshwater available to humanity is shrinking. One of the main causes of this is the polluting of many freshwater resources. In some countries lakes and rivers have become polluted with an assortment of waste, including untreated or partially treated municipal sewage, toxic industrial effluents, Harmful chemicals and ground waters from agricultural activities. Polluted water supplies not only limit water availability but also put millions at risk of water-related diseases.

The lack of freshwater is likely to be one of the most critical natural resource issues facing people in the next 50 years. The world's population is expanding rapidly, yet our supplies of freshwater are not, placing greater demand on our water resources. This makes it even more important that the remaining freshwater we have is kept safe and clean.

The water bodies in present world have been polluted due to escalating population, urbanization and economic activities resulting in severe deterioration of the water quality.

The process of water quality assessment for pollution prediction is a novel application of geo-environmental information. This process requires the assimilation of data which are spatially variable in nature, making geographical information systems (GIS) an ideal tool for such assessments.

A Geographic Information System (GIS) was used to create, combine, and display geographic-data layers from various sources to facilitate interpretation of water-quality data. Geographic data is the information about the earth's surface and features on it. Feeding the data or the information into the system is referred to as 'data capture'.

The use of GIS in water monitoring and management has been long recognized. In fact the capability of this technology offers great tools of how the water quality monitoring and managing can be operationalized in current scenario. Potential application and management is identified in promoting concept of sustainable water resource.

Surface and ground water resources are put to multiple uses in cultivation, manufacturing units, hydro-power, forestry, navigation, domestic uses, aquaculture, recreational-activities. Thus, considering its various attributes, water is indispensable and a unique asset for us.

1.2 Water quality

Water is a universal solvent with a unique characteristic of dissolving a variety of substances. Due to this, large volumes of available surface and ground water resources are able to dissolve a variety of organic, inorganic and toxic wastes generated by the natural processes and anthropogenic activities. This not only deteriorates the water quality but also has an adverse impact on human health impacts of poor water quality is transformed into increased mortality rates, reduced life expectancy, malnutrition, followed by high medical expenditures.

Water plays a vital role in maintaining ecological balance, sustaining soil fertility, development of forest resources and conservation of wildlife. It also plays a pivotal role in all the human developmental activities which has resulted in its pollution and once polluted this unique resource cannot be easily and cost effectively restored to pristine purity.

Water quality is the process to determine the chemical, physical and biological characteristics of water bodies and identifying the source of any possible pollution or contamination which might cause degradation of the water quality.

The domestic's sewage, factories effluents, and agriculture waste can lead to deterioration of water quality. water quality monitoring program are needed in order to raise awareness of public by address the consequences of present and future threats of contamination to water resources.

Water quality involves careful management of both groundwater (recharge areas) and surface water (watersheds, aquifers). Because one sustains the other, cross-contamination is a key concern.

Water quality is a general descriptor of water properties in terms of physical, chemical, thermal, and/or biological characteristics. Agriculture, industrial, and urban areas are anthropogenic sources of point and nonpoint substances. Polluting substances that lead to deterioration of water quality affects most freshwater and estuarine ecosystems in the world.

Point sources are organized sources of pollution with measurable pollution load. Such sources include surface drains transporting waste waters of industries and sewage system, where as non-point sources are non-measurable sources of pollution. Such sources are numerous and influenced by the land-use patterns in the overall watershed. These include both, the natural processes and the human activities.

It is difficult to define a single water quality standard to meet all uses and user needs. For example, physical, chemical, and biological parameters of water that are suitable for human consumption are different from those parameters of water suitable for irrigating a crop.

Water quality is affected by materials delivered to a water body from either point or nonpoint sources. Point sources can be traced to a single source, such as a pipe or a ditch. Nonpoint sources are diffuse and associated with the landscape and its response to water movement, land use and management, and/or other human and natural activities on the watershed.

Monitoring and assessing the quality of surface waters are critical for managing and improving its quality. In situ measurements and collection of water samples for subsequent laboratory analyses are currently used to evaluate.

Variations in availability of water in time, quantity and quality can cause significant fluctuations in the economy of a country. Hence, the conservation, optimum utilization and management of this resource for the betterment of the economic status of the country become paramount. The definition of water quality is very much depending on the desired use of water. Therefore, different uses require different criteria of water quality as well as standard methods for reporting and comparing results of water analysis.

1.3 Geographic Information Systems (GIS)

Geographic Information Systems (GIS) is a computer tool now used to address new and existing water quality problems. GIS uses a computer database to store large quantities of spatial and temporal data. From GIS, spatial distribution mapping for various pollutants can be done. This allows the integration of diverse types of information into a form that makes it possible to consider different approaches to land management and environmental problems before making management decisions.

GIS is very helpful tool for developing solutions for water resources problems to assess in water quality, determining water availability and understanding the natural environment on a broad scale.

The resulting information is very useful for policy makers to take remedial measures.

1.3.1 Basic functions of a GIS

There are two important components of geographic data. The geographic position and attributes or properties related to that position.

Geographic data = Geographic position + attributes/properties

Geographic position gives the location by using a coordinate system.

Attributes refer to properties of spatial entities. Such as,

(i) Identity (town, road, residence etc.),

(ii) Ordinal (ranking such as class 1,2 etc.),

(iii) Scale (elevation, length, width etc.).

Attributes are non-spatial data.

A GIS database will store spatial feature data in a raster or vector format. In the vector format, positions are stored in the form of coordinates. (x,y and sometimes z). A point is described by a single x,y coordinate pair and by its name or label.

Although a line is actually an infinite set of points, in practice a line is described by straight-line segments,

each segment described by a set of coordinate pairs and of course the name or label.

An area also called a polygon, is described by a set of coordinate pairs and by its name or label with the difference that the coordinate pairs at the beginning and the end are the same.

Raster and vector data are used within a GIS framework to produce maps indicating areas of potential hazard to water quality within water resources. Data are also coupled with known modelling techniques to predict and quantify risk frequency and impact.

The potential of GIS to encourage the predictive management of water supply intakes through the assessments of hazard and risk and the modelling of management strategies such as specified grazing areas and the selective use of supply sources. The information science aspects of this development work are of potential interest to end users.

A vector format represents the location and shape of features and boundaries precisely provided they are accurate at the point of input.

Raster data-The grid-based format generalises map features as cells or pixels in a grid matrix. The space is defined by a matrix of points or cells organised into rows and columns. If the rows and columns are numbered, the position of each element can be specified by the column and row numbers.

These can be linked to coordinate positions through the introduction of a coordinate system. Each cell has an attribute value (a number) that represents a geographical phenomenon or nominal data such as elevation, land use class etc. The fineness of the grid or in other words the size of the cells in the grid matrix, will determine the level of detail in which map features can be represented.

Since the past few decades, the increasing of anthropogenic activities especially in industrial area has effects to water bodies. This is the global issues which happening throughout the world. Today, with the advancement of science and technology, the population, industries, agriculture activities, and urban development's have grown up in current world.

A Geographical Information System (GIS) tool was developed and used to construct thematic maps for water quality of the water resources. Environmental data were integrated and an overall picture about the spatial variation in the water quality of the sources can be defined. The water quality maps were derived from the results of previous studies for physico-chemical parameters.

The use of GIS allows people to process and evaluate available data. Without this type of computer tool, such large amounts of data would overwhelm us. Information stored in a GIS will come from a different type of sources. The greater the quantity and quality of the information, the more complete the GIS database will be Sources of information include aerial photographs, satellite images.

Data is entered in the GIS database and the software builds a map of the area. Actually, the computer builds what are called layers—separate maps of the same area, each of which contains different sets of information or themes. These maps are built by providing the computer with geographic coordinates identifying the location of various features.

The monitoring and assessment may be useful for research and policy making purposes. In situ measurements and collection of water samples for subsequent laboratory analyses are currently used to evaluate water quality. These measurements are accurate for a point in time and space but do not give either the spatial or temporal view of water quality in wide space. Thus, the technologies such as GIS are very useful as a tool in evaluating and monitoring water quality.

The techniques GIS-based water quality assessment system to aid decision making in the catchment areas.

1.3.2 Water quality Hazard Mapping

The first stage of water quality assessment of risk is to identify the potential hazard presented by the existing conditions (or potential changes to these conditions). In a catchment area this involves identifying the location of the factors influencing raw water quality at intake site.

1.3.3 Identifying contamination Sources Using Logic Trees

Logic trees are used to combine the physico-chemical characteristics that result in potential pollution sources and therefore predict their spatial distribution using GIS (i.e. they form the basis of the creation of hazard maps).

Fault trees were used in the identification of causal activities in the catchment that may result in an undesirable concentration of a particular substance reaching the intake. Event trees can then be used as a method of assessing the significance of the consequences of such concentrations at the supply intake.

The terms Source Trees and Consequence Trees respectively, are more applicable to their use here. Using the terminology commonly applied to these tree structures, the root of the diagram is determine the catchment information or GIS data needs for the Source Trees and possible outcomes for the Consequence Trees.

The use of such Source Trees combined with hazard mapping is the primary method of hazard identification in the methodology.

1.4 Water Quality and GIS

Factors contributing to the degradation of water quality both natural processes as well as anthropogenic activities adversely affect the quality of water. The major natural sources are the salts introduced from contact of the surface water with various rocks and soil minerals, vegetation, erosion in various stages of bio degradation, dissolved gases native to the environment and volcanic activities. GIS along with in situ water sampling and analysis can simplify and accelerate the procedure for water quality assessment with an acceptable degree of accuracy.

GIS, in being effective tools for monitoring water quality Geographic Information Systems (GIS) that can be used in the management of natural resources. Water quality data incorporation into GIS to evaluate land use impacts on water quality also illustrate how water chemistry data and GIS can be used to examine the impact of land use on water quality, especially with respect to nonpoint source (NPS) pollution at the watershed level.

The GIS maps showed not only contaminant distribution but also illustrated the need to improve the water quality management methods. Using the developed GIS water quality monitoring tool, water quality specialists can concentrate on analyzing data and presenting their results without bothering about the details of the software application being used.

Various studies have been reported on the usefulness of GIS as tools in monitoring of water quality. The use of decision support system like the Geographical Information System (GIS) provides as efficient tool for storing, manipulating and analysis data.

This allows for unmatched monitoring capability to strengthen insights in designing mitigation and management strategies for water quality and water quantity.

A GIS database is used to track actual changes taking place over time within a watershed. Aerial photographs and other spatial data generated at different points in time are used to measure changes in land use, land cover, density of roads, houses, presence or absence of buffer zones around streams, and other items of interest. As this information is collected and evaluated, it is possible to make connections between water quality data (if available) and the changes in land use that have occurred in an area.

Water quality is affected by a variety of natural and anthropogenic factors and as the data are very diverse, the spatial analysis of these data and other factors is a time consuming and complicated task when performed manually.

GIS can be used to calculate loads to a surface water body or to monitor water quality changes within a water body such as a river or bay. A load is the product of flow and concentration, GIS can be used to calculate loads to a surface water body or to monitor water quality changes within a water body such as a and it refers to how much mass of a chemical enters a system in a specified amount of time.

GIS enhance and speed up the various aspects of water resource management such as:

- Water quality assessment,
- Faster and accurate identification of pollution source,
- Location of hot spots of poor water quality on a geographic basis,
- Formulation of mitigation policies and implementation and,
- Feedback on the outcome of the water resource projects.

This will help the water resource manager in better understanding of water quality monitoring and subsequently developing management programs and action plans to ensure that water quality is protected.

Loads to a water body can result from point sources such as industrial discharges or nonpoint sources such as agricultural runoff. Once the loads to a water body are known, water quality models can be used to determine concentration changes within the water body.

Procedures that utilize a GIS have been developed for both types of load calculations and for water quality models.

To properly run a GIS you need hardware, software, data, people and policies and procedures.GIS helps identify and map critical areas of land use and reveal trends that affect water quality. Use GIS to monitor water levels, water usage, and watch for trends to avoid overdraft and drawdown. A major barrier to accurate short-term forecasts is the lack of an efficient system for water quality monitoring. Traditional water quality sampling is time-consuming, expensive, and can only be done for small areas. GIS provides a revolutionary technique to monitor water quality repetitively over a large area.

A GIS based framework will help the decision makers to identify the spatial extent

and causes of water quality problems, such as the effects of land-use practices on the water body of the lake.

Information on water quality can be entered into the database, and, if the location of sample sites is known, these data can be linked to specific locations. This allows the GIS user to make connections between the various physical properties and the quality of water within the area of interest.

1.5 Importance of Geographical Information System

GIS techniques as a promising perspective for effective water quality assessment. Applications of GIS in effective water quality assessment and in taking prompt and consistent water management decisions for the conservation of water resources in current world.

The computer based geographic information system will facilitate the quality analysis of all various types water resources of data. Since GIS are capable of combining large volumes of data from a variety of sources, they are useful tool for many aspects of water quality investigation.

By using a GIS, the natural and man-made influences on water quality in the watersheds were identified and compiled.

In view of the continuously degrading quality of our water resources and its impact on human health, aquatic life and the environment, effective water quality assessment has become critical. These popular techniques however, are limited on temporal and spatial scales of the water quality trends.

The major anthropogenic sources include activities related to land use in agricultural areas, seepage from sewage system and industrial activities. water is confronted by both point and non point source pollutants which contribute to the degradation of its quality.

Limited fresh water resources is a major challenge for the current world. view of the invasive and extensive effects of water management decision on economy, society and the overall environment; relevant, accurate and timely information is a pre-requisite for its strategic management.

The recent advanced techniques like Geographical Information System (GIS) in judicious combination with the conventional in situ measurements are crucial for monitoring and managing the water resources of world.

This technique also provide information on natural resources such crop pattern, land use, land cover, forests, rivers, on a regular basis.

Thus a GIS helps in enhancing the contributions of water quality modelling to practical water quality forecasting, which is essential for sustainable water resources management. GIS techniques, not only their use in water quality monitoring and management will grow but also most of the future water quality models will be the ones with the GIS techniques.

Many researchers have used GIS tools for water quality assessment and have shown many positive results. GIS could also be a powerful tool in developing solutions for water related problems like assessment of water quality, determining availability of water , managing water resources regionally.

GIS was very useful in evaluating and presenting various water quality parameters and also in evaluating and quantifying the impacts of land use changes in the quality of groundwater.

GIS for the assessment of water quality is a superior method as compared with the convectional water quality assessment method. GIS also provide a suitable alternative to the convectional approach.

It is evident from the literature that the GIS techniques are playing a rapidly increasing role in the field of water resource management.

It has long been established that the use of GIS is a resource saving (cost effective), time saving (quick) and more accurate method for water quality monitoring. Geographical Information System techniques have been internationally exploited to gather information needed to monitor various water bodies across the world also allows continuous surveillance of the water bodies helpful in fast detection of changes and trends in the key indicators of water quality.

The assessment of water pollution risks to a surface water intake involves the assimilation of large volumes of spatially variable data. The intrinsic capacity of geographical information systems (GIS) to store, analyse and display such data makes them ideal tools for assisting risk assessments

Geographically large country like India, china, Russia etc. with vast water resources and limited financial assets, could open new and unmatched dimensions in monitoring capabilities of their water resources. It would enable the concerned water regulatory agencies in effective and efficient water quality predictions and in taking prompt and consistent water management decisions for the conservation of the available water resources in the that countries.

ABOUT THE AUTHOR

Dr. Hemant Pathak held positions as Assistant Professor in the department of chemistry, Govt. Indira Gandhi Engineering College, Sagar, MP, India. He had extensive experience in teaching, research and administrative management.

Dr. Pathak received his Ph.D. degree in chemistry from Dr. Hari Singh Gour Central University, Sagar, India and M.Sc. Gold medalist from Jiwaji University, Gwalior. He has published 05 books and more than 50 research papers in reputed International and National journals and received several awards. He is a member of editorial boards of several international journals and societies. His area of specialization includes Engineering Chemistry and Environmental Pollution management.(email- hemantp1981@yahoo.co.in)

www.ingramcontent.com/pod-product-compliance
Lightning Source LLC
Chambersburg PA
CBHW081250170526
45165CB00009B/3272